Fundamentals of Geomechanical and Geotechnical Finite Element Modeling using Abaqus and Python

Joe Cse

Contents

Preface

This book describes the fundamentals of geomechanical and geotechnical finite element modeling using Abaqus and Python. Of particular importance is the probing of non-linear material response of standard soil and rock material models, namely, the Drucker-Prager, Mohr-Coulomb and Cam-Clay models, under triaxial loading. Slope stability and consolidation problems are examined.

Abaqus input (*.inp) files, Python Abaqus post processing and plotting scripts are provided. Python is used because of its wide popularity and is integrated with Abaqus. This enables analysis and post-processing automation, as well as extending the analysis capabilities of Abaqus (for e.g., implementing the strength reduction method for slope stability analysis).

The content of this book is relevant to students, researchers and engineers who are working in civil, energy, and mining industries.

1 Preliminaries

1.1 Stress Invariants

In the following, we use Einstein's summation convention for repeated indices.

- Some invariants:

$$I_1 = \sigma_{kk} \tag{1}$$

$$J_2 = \frac{1}{2} s_{ij} s_{ij} \tag{2}$$

$$J_3 = \frac{1}{3} s_{ij} s_{jk} s_{kl} \tag{3}$$

- σ_{ij} is the Cauchy stress tensor

- s_{ij} is the deviatoric stress tensor:

$$s_{ij} = \sigma_{ij} + \frac{1}{3} p\, \delta_{ij} \tag{4}$$

It can be shown that $s_{11} + s_{22} + s_{33} = 0$.

- δ_{ij} is the identity tensor:

$$\delta_{ij} = \begin{cases} 1, & \text{if i = j} \\ 0, & \text{otherwise} \end{cases} \tag{5}$$

- Mean stress or pressure (sign convention: positive for compression):

$$p = -\frac{1}{3} I_1 = -\frac{1}{3} \sigma_{kk} \tag{6}$$

1

- The (von Mises) shear stress is:

$$q \;=\; \sqrt{3J_2} = \sqrt{\frac{3}{2}s_{ij}s_{ij}} \tag{7}$$

- Lode angle Θ, defined by:

$$\cos(3\Theta) = \left(\frac{r}{q}\right)^3 \tag{8}$$

  where

$$r \;=\; \left(\frac{9}{2}\mathbf{s}\cdot\mathbf{s}:\mathbf{s}\right)^{\frac{1}{3}} = \left[\frac{9}{2}\mathrm{tr}\left(\mathbf{s}^3\right)\right]^{\frac{1}{3}} \tag{9}$$

- Effective pressure p' (sign convention: positive for compression) is

$$p' \;=\; p - u \tag{10}$$

  where u is the pore fluid pressure.

One can use either (I_1, J_2, J_3) or (q, p, Θ) as the set of stress invariants.

2 Triaxial Loading

Probably the most important loading protocol is the triaxial loading, which is typically applied in triaxial testing of soil or rock samples [1]. Triaxial tests are typically used to characterize material properties of soils and rocks. We will use this loading protocol to illustrate common material response plots obtained from triaxial tests. In this chapter and the next, we will primarily focus on the drained condition where fluid pore pressures are not considered. In this case, *effective* stresses experienced by the "soil skeleton" and *total* stresses due to externally applied loads are equal. In Section 5, we will discuss the effects of fluid pore pressures.

The Abaqus input file implementing the triaxial loading protocol is given in Section 2.2. The example setup in the input file consists of:

- A single solid element with side dimensions of 1.0 mm, with symmetry boundary conditions on the negative X, Y and Z sides.

- Material is elastic with elastic modulus of $E = 200 \times 10^3$ kPa and Poisson's ratio of $\nu = 0.2$.

- First analysis step is an initial confinement step in which an isotropic compression of 100 kPa is applied to the element. Second analysis step is the triaxial loading step in which the top surface nodes are displaced downward (Z is vertical direction) while the lateral confinement is maintained at 100 kPa.

3

Running the input file and Python scripts in the following sections, we obtain the response shown in Figure 1.

2.1   Stress Initialization

In the above, the initial confinement is obtained by applying stresses on the sides of the cube element. Therefore, the element will deform due to elastic volumetric changes, and the mean stress or pressure will increase. Offsets on the stresses p', q, and strains ϵ_a, ϵ_v are performed so that the stress and strains quantities in the plots are with respect to the isotropically compressed state. In particular, these quantities are initially zero with respect to this state.

In lieu of this stress initialization approach, a geostatic step can be used in which internal stresses are equilibrated against external applied loading, resulting in (almost) zero initial deformations. We will use the geostatic approach in the context of the Cam Clay model in Section 3.3.

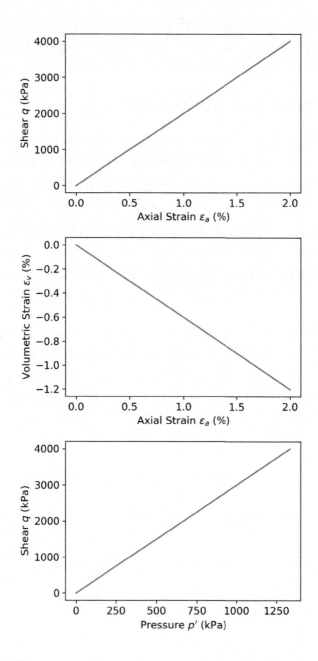

Figure 1: Stress and strain quantities from triaxial test.

2.2 Input File

Save the following lines as 1ElementTriaxial.inp, and then on the command prompt type:

abaqus job=1ElementTriaxial interactive

```
*NODE, NSET=AllNodes
1, 0., 0., 0.
2, 1., 0., 0.
3, 1., 1., 0.
4, 0., 1., 0.
5, 0., 0., 1.
6, 1., 0., 1.
7, 1., 1., 1.
8, 0., 1., 1.
*ELEMENT, TYPE=C3D8, ELSET=PSolid
1, 1, 2, 3, 4, 5, 6, 7, 8
*NSET,NSET=PosXNodes
2,3,6,7
*NSET,NSET=PosYNodes
3,4,7,8
*NSET,NSET=PosZNodes
5,6,7,8
**************************************************
*SURFACE, NAME=PosZSurf, TYPE=ELEMENT
 1, S2
*SURFACE, NAME=PosXSurf, TYPE=ELEMENT
 1, S4
*SURFACE, NAME=PosYSurf, TYPE=ELEMENT
 1, S5
**************************************************
*SOLID SECTION, ELSET=PSolid, MATERIAL=MSolid
**************************************************
***
*** Units:
***    mass   = kg
***    length = mm
***    time   = s
```

6

```
***    force = mN
***    stress = 1.0e+03 Pa
***
*PARAMETER
Eyoung  = 200.e3
nu = 0.2
***************
ConfinementPressure = 100.
***************
COMP_STRAIN = 0.1
STEP_VELOC = -COMP_STRAIN
*************
*MATERIAL, NAME=MSolid
*ELASTIC
<Eyoung>,<nu>
************************************************
** Load Amplitudes
************************************************
*AMPLITUDE, TIME=TOTAL TIME, NAME=ConfinementPressure_Amplitude
0.0,0.0
1.0,<ConfinementPressure>
2.0,<ConfinementPressure>
************************************************
** Step 1 - Initial Confinement
************************************************
*STEP, UNSYMM=YES, NLGEOM=NO, INC=999
Confinement
*STATIC
0.01,1.0,1.e-12,0.1
**********************
** Boundary conditions
**********************
*BOUNDARY, OP=MOD
1, 1
1, 2
1, 3
2, 2
2, 3
3, 3
4, 1
4, 3
5, 1
```

```
5, 2
6, 2
8, 1
***********************
** Confinement pressure
***********************
*DSLOAD, AMPLITUDE=ConfinementPressure_Amplitude, OP=NEW
PosXSurf,P,1.0
PosYSurf,P,1.0
PosZSurf,P,1.0
*******************
** ODB output
*******************
*OUTPUT, FIELD, FREQUENCY=1
*NODE OUTPUT
U,RF
*ELEMENT OUTPUT
E,S
*******************
** History output
*******************
*OUTPUT, HISTORY, FREQUENCY=1
*ELEMENT OUTPUT, ELSET=PSolid
E11,E22,E33,S11,S22,S33,MISES,PRESS
*END STEP
**************************************************
** Step 2 - Triaxial Loading
**************************************************
*STEP, UNSYMM=YES, NLGEOM=NO, INC=99999999
Triaxial Loading
*STATIC
0.01,1.0,1.e-12,0.1
***********************
** Boundary conditions
***********************
*BOUNDARY, TYPE=VELOCITY, OP=NEW
1, 1
1, 2
1, 3
2, 2
2, 3
3, 3
```

```
4, 1
4, 3
5, 1
5, 2
6, 2
8, 1
PosZNodes,3,3,<STEP_VELOC>
***********************
** Confinement pressure
***********************
*DSLOAD, AMPLITUDE=ConfinementPressure_Amplitude, OP=NEW
PosXSurf,P,1.0
PosYSurf,P,1.0
*END STEP
```

2.3   Abaqus Python Postprocessing Script

Save the following lines as **postprocess.py**, and then on the command prompt type:

abaqus python postprocess.py

This script needs only slight modification to postprocess the material models in the next chapter.

```python
from odbAccess import *
from odbMaterial import *
from odbSection import *
from abaqusConstants import *
import numpy as np

odbName='1Element_Triaxial.odb'

odb = openOdb(odbName)
assembly = odb.rootAssembly

elementSet = \
    odb.rootAssembly.instances['PART-1-1'].elementSets['PSOLID']

axial_strain      = []
volumetric_strain = []
Pvec              = []
Qvec              = []

print('Processing Step: %d\n'% odb.steps['Step-2'].number)

index=0
integ_point=0

for frame in odb.steps['Step-2'].frames:

    print('Frame: %d'% index)

    StrainSet = frame.fieldOutputs['E']
```

10

```python
StressSet = frame.fieldOutputs['S']

E1 = StrainSet.values[integ_point].data[0] # E11
E2 = StrainSet.values[integ_point].data[1] # E22
E3 = StrainSet.values[integ_point].data[2] # E33

EV = E1+E2+E3

S1 = StressSet.values[integ_point].data[0] # S11
S2 = StressSet.values[integ_point].data[1] # S22
S3 = StressSet.values[integ_point].data[2] # S33

#
# Convention:
#   compression is positive
#
P = -(S1+S2+S3)/3.0
Q = sqrt( 0.5*( (S1-S2)**2+(S2-S3)**2+(S3-S1)**2 ) )

#
# Convention:
#   axial strain (percent) is positive in compression
#   volumetric strain (percent) is positive for dilation
#
axial_strain.append(-E3*100)
volumetric_strain.append(EV*100)
Pvec.append(P)
Qvec.append(Q)
index=index+1
# end for

# Close the output database before exiting the program
odb.close()

numdata = len(axial_strain)

f = open('results.csv', 'w')
f.write('Axial Strain (%), Volumetric Strain (%),\
    Shear Q, Pressure P\n')
for i in range(0,numdata):
f.write('%20.8e,%20.8e,%20.8e,%20.8e\n'\
        %(axial_strain[i],volumetric_strain[i],Qvec[i],Pvec[i]))
```

```
# end for

f.close()
```

2.4 Plotting script

Note: the following script requires a separate Python installation that includes the Matplotlib plotting library. Save the following lines as `plot_results.py`, and then on the command prompt type:

`python plot_results.py`

```python
import matplotlib.pyplot as plt
plt.rcParams.update({'font.size': 12})

fin = open('results.csv','r')
lines = fin.readlines()
fin.close()

axial_strain      = []
volumetric_strain = []
Pvec              = []
Qvec              = []
for i,line in enumerate(lines):
    if i > 0: # skip headers
        print(line)
        line = line.strip()
        line = line.split(',')
        EA = float(line[0])
        EV = float(line[1])
        Q  = float(line[2])
        P  = float(line[3])
        axial_strain.append(EA)
        volumetric_strain.append(EV)
        Pvec.append(P)
        Qvec.append(Q)
    # end if
# end for

fig = plt.figure(figsize=(5,10))
fig.subplots_adjust(hspace=0.5)
```

```python
ax1 = fig.add_subplot(3,1,1)

#
# Perform offset
#
def voffset(arr):
    v0  = arr[0]
    arr = [val - v0 for val in arr]
    return arr
# end def

Qvec = voffset(Qvec)
Pvec = voffset(Pvec)
axial_strain = voffset(axial_strain)
volumetric_strain = voffset(volumetric_strain)


ax1.plot(axial_strain,Qvec)
ax1.set_xlabel('Axial Strain $\epsilon_a$ (%)')
ax1.set_ylabel('Shear $q$ (kPa)')

ax2 = fig.add_subplot(3,1,2)
ax2.plot(axial_strain,volumetric_strain)
ax2.set_xlabel('Axial Strain $\epsilon_a$ (%)')
ax2.set_ylabel('Volumetric Strain $\epsilon_v$ (%)')

ax3 = fig.add_subplot(3,1,3)
ax3.plot(Pvec,Qvec)
ax3.set_xlabel('Pressure $p\'$ (kPa)')
ax3.set_ylabel('Shear $q$ (kPa)')

plt.tight_layout()
plt.savefig('qp_plot.png', dpi=300)
plt.show()
```

14

3   Material Models

We discuss the following material models:

- Drucker-Prager model

- Mohr-Coulomb model

- Cam-Clay model

In the following, we assume the drained condition where the pore fluid pressure $u = 0$. Therefore, the effective pressure is equal to the total pressure $p' = p$.

3.1 Drucker-Prager model

The Drucker-Prager model is a prototypical soil or rock constitutive model that has a pressure-dependent yield stress. There are other variants of the Drucker-Prager model in Abaqus, but for simplicity, we will describe the so-called linear Drucker-Prager model. Its yield function is given by:

$$F = q - (p' \tan \phi + c) \tag{11}$$

where ϕ is the friction angle. This equation means that the material yields when $F = 0$, in which the shear stress q will be:

$$q = p' \tan \phi + c \tag{12}$$

As can be seen, the yield stress depends on the effective mean stress or pressure p'. We can illustrate the dependency of the yield stress by performing the triaxial test with several levels of confinement as shown below.

Another important function is the so-called plastic potential function

$$G = q - p' \tan \psi \tag{13}$$

The incremental plastic strains are determined through the so-called flow rule:

$$\Delta \epsilon_{ij}^p = \Delta \lambda \frac{\partial G}{\partial \sigma_{ij}} \tag{14}$$

16

where $\Delta\lambda$ is a value that is determined internally by Abaqus, and the derivative of the plastic potential can be shown to be:

$$\frac{\partial G}{\partial \sigma_{ij}} = \frac{3}{2q}s_{ij} - \frac{\tan\psi}{3}\delta_{ij} \tag{15}$$

The above may be difficult to understand without background of continuum mechanics, but what one needs to essentially know is that the plastic strain increment is proportional to $\frac{\partial G}{\partial \sigma_{ij}}$:

$$\Delta\epsilon_{ij}^p \propto \frac{\partial G}{\partial \sigma_{ij}} \tag{16}$$

and that the volumetric part of the plastic strain (dilation or volume expansion is taken as positive) is:

$$
\begin{aligned}
\Delta\epsilon_v^p &= -\left(\Delta\epsilon_{11}^p + \Delta\epsilon_{22}^p + \Delta\epsilon_{33}^p\right) \\
&\propto \frac{\tan\psi}{3}\left(\delta_{11} + \delta_{22} + \delta_{33}\right) \\
&= \frac{\tan\psi}{3}3 \\
&= \tan\psi
\end{aligned}
\tag{17}
$$

Therefore, if the dilation angle ψ is zero, the volumetric plastic strain increment will be zero (note: there will still be volume change due to elasticity). This means that after yield, the volumetric strain should remain constant.

Consider the material model given by the input file in Section 3.1.1. The material properties are:

- Elastic properties: elastic modulus $E = 200 \times 10^3$ kPa, Poisson's ratio $\nu = 0.2$

- Plastic properties: friction angle $\beta = 30°$, dilation angle $\psi = 0°$, cohesion $c = 1 \times 10^3$ kPa

Running this model, we obtain the response shown in Figure 2. As can be seen, the volumetric strain stays constant after the material has yielded, and the material remains in a contracted state.

If we rerun the above model with the dilation angle set to $\psi = \beta = 30°$, we get the response shown in Figure 3. One sees that the volumetric strain changes the material from a contracted to a dilated state.

The dilation angle is generally different (lower) than the friction angle, which results in an unsymmetric tangent moduli matrix. Therefore, the unsymmetric solver needs to be used in the analysis step by adding the option UNSYMM=YES in the keyword *STEP.

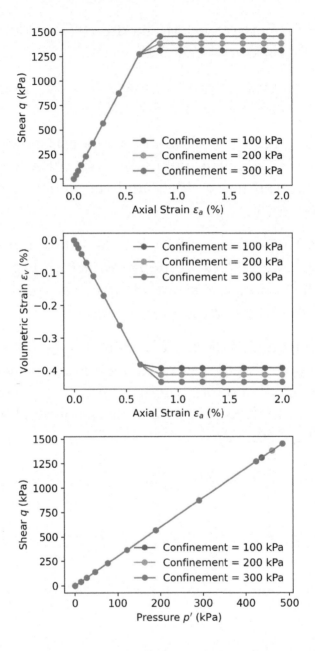

Figure 2: Response of Drucker-Prager model with dilation angle $\psi = 0°$.

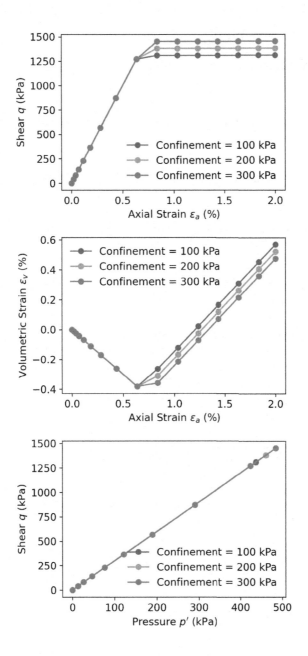

Figure 3: Response of Drucker-Prager model with dilation angle $\psi = 30°$.

3.1.1   Drucker-Prager Input File

```
*NODE, NSET=AllNodes
1, 0., 0., 0.
2, 1., 0., 0.
3, 1., 1., 0.
4, 0., 1., 0.
5, 0., 0., 1.
6, 1., 0., 1.
7, 1., 1., 1.
8, 0., 1., 1.
*ELEMENT, TYPE=C3D8, ELSET=PSolid
1, 1, 2, 3, 4, 5, 6, 7, 8
*NSET,NSET=PosXNodes
2,3,6,7
*NSET,NSET=PosYNodes
3,4,7,8
*NSET,NSET=PosZNodes
5,6,7,8
**************************************************
*SURFACE, NAME=PosZSurf, TYPE=ELEMENT
 1, S2
*SURFACE, NAME=PosXSurf, TYPE=ELEMENT
 1, S4
*SURFACE, NAME=PosYSurf, TYPE=ELEMENT
 1, S5
**************************************************
*SOLID SECTION, ELSET=PSolid, MATERIAL=MSolid
**************************************************
***
*** Units:
***    mass   = kg
***    length = mm
***    time   = s
***    force  = mN
***    stress = 1.0e+03 Pa
***
*PARAMETER
Eyoung  = 200.e3
nu = 0.2
phi = 30.
psi = 30.
```

```
coh = 1.e3
**************
ConfinementPressure = 100.
**************
COMP_STRAIN = 0.02
STEP_VELOC = -COMP_STRAIN
*************
*MATERIAL, NAME=MSolid
*ELASTIC
<Eyoung>,<nu>
*DRUCKER PRAGER, SHEAR CRITERION=LINEAR
<phi>,,<psi>
*DRUCKER PRAGER HARDENING, TYPE=SHEAR
<coh>
***********************************************
** Load Amplitudes
***********************************************
*AMPLITUDE, TIME=TOTAL TIME, NAME=ConfinementPressure_Amplitude
0.0,0.0
1.0,<ConfinementPressure>
2.0,<ConfinementPressure>
***********************************************
** Step 1 - Initial Confinement
***********************************************
*STEP, UNSYMM=YES, NLGEOM=NO, INC=999
Confinement
*STATIC
0.01,1.0,1.e-12,0.1
***********************
** Boundary conditions
***********************
*BOUNDARY, OP=MOD
1, 1
1, 2
1, 3
2, 2
2, 3
3, 3
4, 1
4, 3
5, 1
5, 2
```

```
6, 2
8, 1
***********************
** Confinement pressure
***********************
*DSLOAD, AMPLITUDE=ConfinementPressure_Amplitude, OP=NEW
PosXSurf,P,1.0
PosYSurf,P,1.0
PosZSurf,P,1.0
*******************
** ODB output
*******************
*OUTPUT, FIELD, FREQUENCY=1
*NODE OUTPUT
U,RF
*ELEMENT OUTPUT
E,S
*******************
** History output
*******************
*OUTPUT, HISTORY, FREQUENCY=1
*ELEMENT OUTPUT, ELSET=PSolid
E11,E22,E33,S11,S22,S33,MISES,PRESS
*END STEP
************************************************
** Step 2 - Triaxial Loading
************************************************
*STEP, UNSYMM=YES, NLGEOM=NO, INC=99999999
Triaxial Loading
*STATIC
0.01,1.0,1.e-12,0.1
***********************
** Boundary conditions
***********************
*BOUNDARY, TYPE=VELOCITY, OP=NEW
1, 1
1, 2
1, 3
2, 2
2, 3
3, 3
4, 1
```

23

```
4, 3
5, 1
5, 2
6, 2
8, 1
PosZNodes,3,3,<STEP_VELOC>
***********************
** Confinement pressure
***********************
*DSLOAD, AMPLITUDE=ConfinementPressure_Amplitude, OP=NEW
PosXSurf,P,1.0
PosYSurf,P,1.0
*END STEP
```

3.1.2 Drucker-Prager Plotting script

```
import matplotlib.pyplot as plt
plt.rcParams.update({'font.size': 12})

confinementVec = [100,200,300]
figfilename = 'qp_plotDP_dilation.png'

fig = plt.figure(figsize=(5,10))
fig.subplots_adjust(hspace=0.5)
ax1 = fig.add_subplot(3,1,1)
ax2 = fig.add_subplot(3,1,2)
ax3 = fig.add_subplot(3,1,3)

def voffset(arr):
    v0  = arr[0]
    arr = [val - v0 for val in arr]
    return arr
# end def

for confinement in confinementVec:

    confinement = round(confinement)
    jobName = '1Element_DPTriaxial_' + str(confinement)
    fin = open('results_'+jobName+'.csv','r')
    lines = fin.readlines()
    fin.close()

    axial_strain      = []
    volumetric_strain = []
    Pvec              = []
    Qvec              = []

    for i,line in enumerate(lines):
        if i > 0: # skip headers
            print(line)
            line = line.strip()
            line = line.split(',')
            EA = float(line[0])
            EV = float(line[1])
            Q  = float(line[2])
            P  = float(line[3])
```

```
        axial_strain.append(EA)
        volumetric_strain.append(EV)
        Pvec.append(P)
        Qvec.append(Q)
    # end if
# end for

#
# Perform offset
#
Qvec = voffset(Qvec)
Pvec = voffset(Pvec)
axial_strain = voffset(axial_strain)
volumetric_strain = voffset(volumetric_strain)

ax1.plot(axial_strain,Qvec,'-o',\
    label='Confinement = %d kPa'%(confinement))

ax2.plot(axial_strain,volumetric_strain,'-o',\
    label='Confinement = %d kPa'%(confinement))

ax3.plot(Pvec,Qvec,'-o',\
    label='Confinement = %d kPa'%(confinement))


# end for
ax1.legend(framealpha=0.0)
ax2.legend(framealpha=0.0, loc='upper left')
ax3.legend(framealpha=0.0, loc='lower right')

ax1.set_xlabel('Axial Strain $\epsilon_a$ (%)')
ax1.set_ylabel('Shear $q$ (kPa)')
ax2.set_xlabel('Axial Strain $\epsilon_a$ (%)')
ax2.set_ylabel('Volumetric Strain $\epsilon_v$ (%)')
ax3.set_xlabel('Pressure $p\'$ (kPa)')
ax3.set_ylabel('Shear $q$ (kPa)')

plt.tight_layout()
plt.savefig(figfilename, dpi=300)
plt.show()
```

3.2 Mohr-Coulomb model

The Mohr-Coulomb model adds complexity to the Drucker-Prager model by making the yield stress dependent on another (third) invariant, in addition to shear q and pressure p'. The yield function of the Mohr-Coulomb model is:

$$F = R_{mc}\, q - (p' \tan \phi + c) \qquad (18)$$

which is similar in form to the Drucker-Prager, except that it has the additional term R_{mc} that multiplies the shear q:

$$
\begin{aligned}
R_{mc}(\Theta, \phi) &= \frac{1}{\sqrt{3}\cos\phi} \sin\left(\Theta + \frac{\pi}{3}\right) \\
&\quad + \frac{1}{3}\cos\left(\Theta + \frac{\pi}{3}\right)\tan\phi \qquad (19)
\end{aligned}
$$

where the third invariant is Θ, which is the Lode angle defined in (8) and (9).

On the compression meridian, $\Theta = \frac{\pi}{3}$, and on the extension meridian, $\Theta = 0$. This means that the yield stresses on the compression and extension (squeezing) meridians are generally different, as will be shown in the example below.

The plastic potential is geometrically similar to the yield function, but is a function of the dilation angle ψ instead of the friction angle. Therefore, the incremental plastic strain will additionally depend on the third invariant r. In Abaqus, the sharp corners of the plastic potential have been smoothed to prevent numerical singularities.

Consider the material model used in the input files in Sections 3.2.1 and 3.2.2. The material properties are:

- Elastic properties: elastic modulus $E = 200 \times 10^3$ kPa, Poisson's ratio $\nu = 0.2$

- Plastic properties: friction angle $\phi = 30°$, dilation angle $\psi = 0.1°$ (minimum value in Abaqus), cohesion $c = 1 \times 10^3$ kPa

Under triaxial compression and extension with an initial confinement of 100 kPa, we obtain the material response shown in Figure 4. In the last subfigure, the compression and extension yield functions (F_{comp} and F_{ext}, respectively) have been plotted.

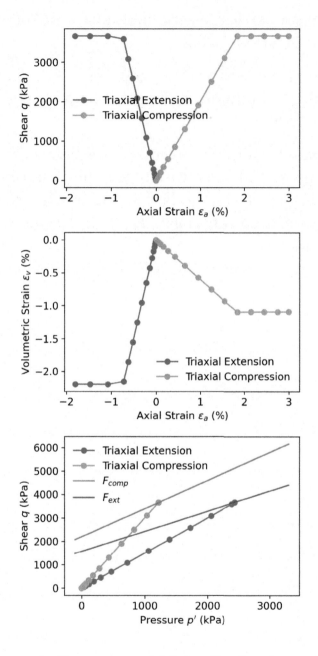

Figure 4: Response of Mohr-Coulomb model.

3.2.1 Mohr-Coulomb Triaxial Compression Input File

```
*NODE, NSET=AllNodes
1, 0., 0., 0.
2, 1., 0., 0.
3, 1., 1., 0.
4, 0., 1., 0.
5, 0., 0., 1.
6, 1., 0., 1.
7, 1., 1., 1.
8, 0., 1., 1.
*ELEMENT, TYPE=C3D8, ELSET=PSolid
1, 1, 2, 3, 4, 5, 6, 7, 8
*NSET,NSET=PosXNodes
2,3,6,7
*NSET,NSET=PosYNodes
3,4,7,8
*NSET,NSET=PosZNodes
5,6,7,8
*************************************************
*SURFACE, NAME=PosZSurf, TYPE=ELEMENT
 1, S2
*SURFACE, NAME=PosXSurf, TYPE=ELEMENT
 1, S4
*SURFACE, NAME=PosYSurf, TYPE=ELEMENT
 1, S5
*************************************************
*SOLID SECTION, ELSET=PSolid, MATERIAL=MSolid
*************************************************
***
*** Units:
***    mass   = kg
***    length = mm
***    time   = s
***    force  = mN
***    stress = 1.0e+03 Pa
***
*PARAMETER
Eyoung  = 200.e3
nu = 0.2
phi = 30.
```

```
psi = 0.1
coh = 1.e3
**************
ConfinementPressure = 100.
**************
COMP_STRAIN = 0.03
STEP_VELOC = -COMP_STRAIN
*************
*MATERIAL, NAME=MSolid
*ELASTIC
<Eyoung>,<nu>
*MOHR COULOMB
<phi>,<psi>
*MOHR COULOMB HARDENING
<coh>
***************************************************
** Load Amplitudes
***************************************************
*AMPLITUDE, TIME=TOTAL TIME, NAME=ConfinementPressure_Amplitude
0.0,0.0
1.0,<ConfinementPressure>
2.0,<ConfinementPressure>
***************************************************
** Step 1 - Initial Confinement
***************************************************
*STEP, UNSYMM=YES, NLGEOM=NO, INC=999
Confinement
*STATIC
0.01,1.0,1.e-12,0.1
**********************
** Boundary conditions
**********************
*BOUNDARY, OP=MOD
1, 1
1, 2
1, 3
2, 2
2, 3
3, 3
4, 1
4, 3
5, 1
```

31

```
5, 2
6, 2
8, 1
***********************
** Confinement pressure
***********************
*DSLOAD, AMPLITUDE=ConfinementPressure_Amplitude, OP=NEW
PosXSurf,P,1.0
PosYSurf,P,1.0
PosZSurf,P,1.0
********************
** ODB output
********************
*OUTPUT, FIELD, FREQUENCY=1
*NODE OUTPUT
U,RF
*ELEMENT OUTPUT
E,S
********************
** History output
********************
*OUTPUT, HISTORY, FREQUENCY=1
*ELEMENT OUTPUT, ELSET=PSolid
E11,E22,E33,S11,S22,S33,MISES,PRESS
*END STEP
*************************************************
** Step 2 - Triaxial Loading
*************************************************
*STEP, UNSYMM=YES, NLGEOM=NO, INC=99999999
Triaxial Compression
*STATIC
0.01,1.0,1.e-12,0.1
***********************
** Boundary conditions
***********************
*BOUNDARY, TYPE=VELOCITY, OP=NEW
1, 1
1, 2
1, 3
2, 2
2, 3
3, 3
```

```
4, 1
4, 3
5, 1
5, 2
6, 2
8, 1
PosZNodes,3,3,<STEP_VELOC>
************************
** Confinement pressure
************************
*DSLOAD, AMPLITUDE=ConfinementPressure_Amplitude, OP=NEW
PosXSurf,P,1.0
PosYSurf,P,1.0
*END STEP
```

3.2.2 Mohr-Coulomb Triaxial Extension Input File

```
*NODE, NSET=AllNodes
1, 0., 0., 0.
2, 1., 0., 0.
3, 1., 1., 0.
4, 0., 1., 0.
5, 0., 0., 1.
6, 1., 0., 1.
7, 1., 1., 1.
8, 0., 1., 1.
*ELEMENT, TYPE=C3D8, ELSET=PSolid
1, 1, 2, 3, 4, 5, 6, 7, 8
*NSET,NSET=PosXNodes
2,3,6,7
*NSET,NSET=PosYNodes
3,4,7,8
*NSET,NSET=PosZNodes
5,6,7,8
*************************************************
*SURFACE, NAME=PosZSurf, TYPE=ELEMENT
 1, S2
*SURFACE, NAME=PosXSurf, TYPE=ELEMENT
 1, S4
*SURFACE, NAME=PosYSurf, TYPE=ELEMENT
 1, S5
*************************************************
*SOLID SECTION, ELSET=PSolid, MATERIAL=MSolid
*************************************************
***
*** Units:
***    mass   = kg
***    length = mm
***    time   = s
***    force  = mN
***    stress = 1.0e+03 Pa
***
*PARAMETER
Eyoung  = 200.e3
nu = 0.2
phi = 30.
psi = 0.1
```

34

```
coh = 1.e3
***************
ConfinementPressure = 100.
***************
COMP_STRAIN = 0.02
STEP_VELOC = -COMP_STRAIN
*************
*MATERIAL, NAME=MSolid
*ELASTIC
<Eyoung>,<nu>
*MOHR COULOMB
<phi>,<psi>
*MOHR COULOMB HARDENING
<coh>
************************************************
** Load Amplitudes
************************************************
*AMPLITUDE, TIME=TOTAL TIME, NAME=ConfinementPressure_Amplitude
0.0,0.0
1.0,<ConfinementPressure>
2.0,<ConfinementPressure>
************************************************
** Step 1 - Initial Confinement
************************************************
*STEP, UNSYMM=YES, NLGEOM=NO, INC=999
Confinement
*STATIC
0.01,1.0,1.e-12,0.1
**********************
** Boundary conditions
**********************
*BOUNDARY, OP=MOD
1, 1
1, 2
1, 3
2, 2
2, 3
3, 3
4, 1
4, 3
5, 1
5, 2
```

35

```
6, 2
8, 1
***********************
** Confinement pressure
***********************
*DSLOAD, AMPLITUDE=ConfinementPressure_Amplitude, OP=NEW
PosXSurf,P,1.0
PosYSurf,P,1.0
PosZSurf,P,1.0
*******************
** ODB output
*******************
*OUTPUT, FIELD, FREQUENCY=1
*NODE OUTPUT
U,RF
*ELEMENT OUTPUT
E,S
*******************
** History output
*******************
*OUTPUT, HISTORY, FREQUENCY=1
*ELEMENT OUTPUT, ELSET=PSolid
E11,E22,E33,S11,S22,S33,MISES,PRESS
*END STEP
*************************************************
** Step 2 - Triaxial Loading
*************************************************
*STEP, UNSYMM=YES, NLGEOM=NO, INC=99999999
Triaxial Extension
*STATIC
0.01,1.0,1.e-12,0.1
***********************
** Boundary conditions
***********************
*BOUNDARY, TYPE=VELOCITY, OP=NEW
1, 1
1, 2
1, 3
2, 2
2, 3
3, 3
4, 1
```

```
4, 3
5, 1
5, 2
6, 2
8, 1
PosXNodes,1,1,<STEP_VELOC>
PosYNodes,2,2,<STEP_VELOC>
***********************
** Confinement pressure
***********************
*DSLOAD, AMPLITUDE=ConfinementPressure_Amplitude, OP=NEW
PosZSurf,P,1.0
*END STEP
```

3.2.3 Mohr-Coulomb Plotting script

```python
import matplotlib.pyplot as plt
import math
import numpy as np
plt.rcParams.update({'font.size': 12})

loadVec = ['Extension','Compression']
figfilename = 'qp_plotMC.png'

fig = plt.figure(figsize=(5,10))
fig.subplots_adjust(hspace=0.5)
ax1 = fig.add_subplot(3,1,1)
ax2 = fig.add_subplot(3,1,2)
ax3 = fig.add_subplot(3,1,3)

def voffset(arr):
    v0  = arr[0]
    arr = [val - v0 for val in arr]
    return arr
# end def

for load in loadVec:

    jobName = '1Element_MCTriaxial_' + load
    fin = open('results_'+jobName+'.csv','r')
    lines = fin.readlines()
    fin.close()

    axial_strain      = []
    volumetric_strain = []
    Pvec              = []
    Qvec              = []
    for i,line in enumerate(lines):
        if i > 0: # skip headers
            print(line)
            line = line.strip()
            line = line.split(',')
            EA = float(line[0])
            EV = float(line[1])
            Q  = float(line[2])
            P  = float(line[3])
```

```
            axial_strain.append(EA)
            volumetric_strain.append(EV)
            Pvec.append(P)
            Qvec.append(Q)
        # end if
    # end for

    #
    # Perform offset
    #
    Qvec = voffset(Qvec)
    Pvec = voffset(Pvec)
    axial_strain = voffset(axial_strain)
    volumetric_strain = voffset(volumetric_strain)

    ax1.plot(axial_strain,Qvec,'-o',\
        label='Triaxial ' + load)

    ax2.plot(axial_strain,volumetric_strain,'-o',\
        label='Triaxial ' + load)

    ax3.plot(Pvec,Qvec,'-o',\
        label='Triaxial ' + load)

# end for

#
# plot yield surface
#
def Rmc(ThetaDeg,phideg):

    ThetaRad = ThetaDeg*math.pi/180.
    phirad   = phideg*math.pi/180.

    cos_phi = math.cos(phirad)
    tan_phi = math.tan(phirad)

    sin_phiplus = math.sin(ThetaRad+math.pi/3.)
    cos_phiplus = math.cos(ThetaRad+math.pi/3.)

    afac = sin_phiplus/(math.sqrt(3.)*cos_phi)
    bfac = cos_phiplus*tan_phi/3.
```

```
      return afac+bfac

# end def
P0 = 100.
cohesion = 1.e3
phideg = 30.
phirad = phideg*math.pi/180.
tan_phi = math.tan(phirad)

Rmc_Comp = abs(Rmc(60.,phideg))
Rmc_Ext  = abs(Rmc( 0.,phideg))
print(Rmc_Comp)
print(Rmc_Ext)

F_p = np.arange(0.,3500.,100.)
F_q_comp = (F_p*tan_phi + cohesion)/Rmc_Comp
F_q_ext  = (F_p*tan_phi + cohesion)/Rmc_Ext

#
# Pressure offset
# note: initial confinement P0 is
# the same for both triaxial
# compression and extension
#
F_p = F_p-P0
ax3.plot(F_p,F_q_comp,'-',\
    label='$F_{comp}$')

ax3.plot(F_p,F_q_ext,'-',\
    label='$F_{ext}$')

ax1.legend(framealpha=0.0, loc='center left')
ax2.legend(framealpha=0.0, loc='lower right')
ax3.legend(framealpha=0.0, loc='upper left')

ax1.set_xlabel('Axial Strain $\epsilon_a$ (%)')
ax1.set_ylabel('Shear $q$ (kPa)')
ax2.set_xlabel('Axial Strain $\epsilon_a$ (%)')
ax2.set_ylabel('Volumetric Strain $\epsilon_v$ (%)')
ax3.set_xlabel('Pressure $p\'$ (kPa)')
ax3.set_ylabel('Shear $q$ (kPa)')
```

```
plt.tight_layout()
plt.savefig(figfilename, dpi=300)
plt.show()
```

3.3 Cam-Clay model

The Cam-Clay model is a prototypical soil model for consolidation [1]. We consider a specific form of the Cam-Clay model with the following material parameters:

- Logarithmic elastic bulk modulus (dimensionless) κ

- Logarithmic plastic bulk modulus (dimensionless) λ

- Poisson's ratio ν

- Slope of the so-called critical state line M

- Initial center of the ellipse a_0

- Initial void ratio e_0

The elastic response is characterized by κ and ν, which are specified under the keyword *POROUS ELASTIC. The initial void ratio e_0 is specified as an *initial condition* using the keyword *INITIAL CONDITIONS, TYPE=RATIO, and the remaining parameters are specified under the keyword *CLAY PLASTICITY. The adjective "logarithmic" refers to the use of $\ln p'$ as the stress axis in the stress-strain space; the strain axis is the void ratio e.

The yield function is:

$$F = \left(\frac{p'}{a} - 1\right)^2 + \left(\frac{q}{Ma}\right)^2 - 1 \tag{20}$$

which is an equation of an ellipse, where the current center of the ellipse is at $(p', q) = (a, 0)$. The major radius is a and the minor radius is $M\,a$. M is the slope of the so-called critical state line. Under continuous shearing, the material approaches a well-defined critical state in which the material flows as a frictional fluid. At the onset of the critical state, shear distortions ϵ_s occur without any further changes in mean effective stress p', deviatoric shear stress q or specific volume ν:

$$\frac{\partial p'}{\partial \epsilon_s} = 0 \tag{21}$$

$$\frac{\partial q}{\partial \epsilon_s} = 0 \tag{22}$$

$$\frac{\partial \upsilon}{\partial \epsilon_s} = 0 \tag{23}$$

The plastic potential is taken to be identical to the yield function $G = F$, and contribute to incremental plastic strains the same as that described by (14).

We consider the exponential hardening mechanism for the yield surface:

$$a = a_0 \exp\left[(1 + e_0)\frac{1 - J^{pl}}{\lambda - \kappa J^{pl}}\right] \tag{24}$$

where J^{pl} is the plastic volume change, which is calculated by Abaqus.

The "over-consolidation ratio" (OCR) is defined as the highest pressure experienced divided by the current pressure:

$$\text{OCR} = \frac{p'_{max}}{p'_0} \tag{25}$$

43

where $p'_{max} = 2\,a_0$ and p'_0 is current confinement pressure. A soil that is currently experiencing the highest pressure experienced ($p'_0 = p'_{max}$) is said to be "normally consolidated" and has an OCR $= 1$. A soil that is currently having a confinement pressure lower than the highest pressure experienced ($p'_0 < p'_{max}$) is said to be "overconsolidated" and has an OCR > 1. Depending on the OCR, the material response can undergo either hardening or softening response, as will be shown in the example below. The stress history of a site (through the OCR) is of particular importance for geomechanical and geotechnical problems such as excavations.

In Abaqus, the current mean effective pressure p'_0 is set as an *initial condition* using the keyword

*INITIAL CONDITIONS, TYPE=STRESS.

A geostatic step *GEOSTATIC is used for stress initialization.

Consider the material model used in the input file in Section 3.3.1. The material properties are:

- Logarithmic elastic bulk modulus (dimensionless) $\kappa = 0.0066$

- Logarithmic plastic bulk modulus (dimensionless)$\lambda = 0.077$

- Poisson's ratio $\nu = 0.3$

- Slope of the so-called critical state line $M = 1.2$

- Initial void ratio $e_0 = 0.5$

The following OCRs are considered:

- Normally consolidated, OCR $= 1$: $2\,a_0 = 200$ kPa, $p'_0 = 200$ kPa

- Lightly overconsolidated, OCR $= 2$: $2\,a_0 = 200$ kPa, $p'_0 = 100$ kPa

- Heavily overconsolidated, OCR $= 5$: $2\,a_0 = 500$ kPa, $p'_0 = 100$ kPa

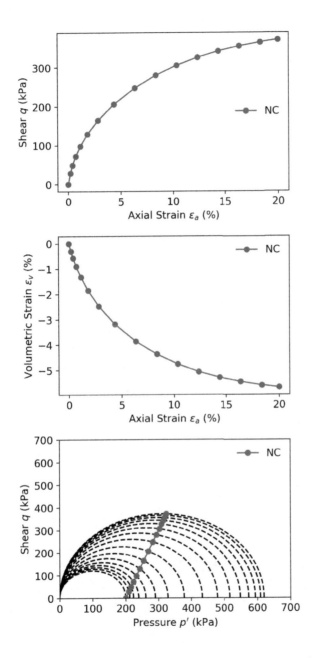

Figure 5: Response of Cam-Clay model - Normally Consolidated.

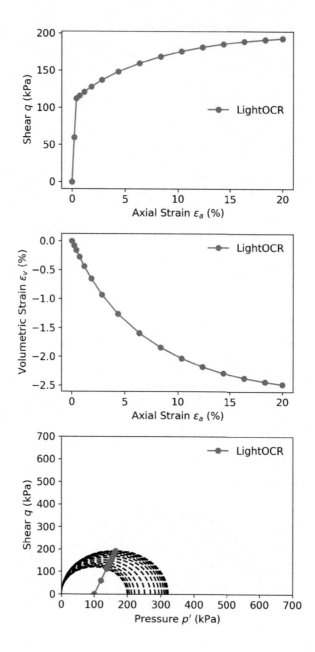

Figure 6: Response of Cam-Clay model - Lightly Consolidated.

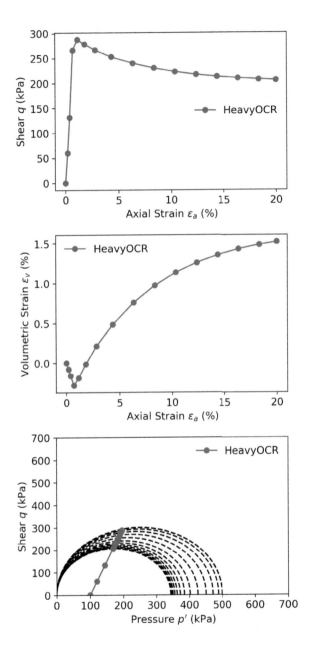

Figure 7: Response of Cam-Clay model - Heavily Consolidated.

3.3.1 Cam-Clay Triaxial Compression Input File

Comment out appropriate lines with:

Pc0, a0, ConfinementPressure

to get the normally consolidated, lightly overconsolidated and heavily overconsolidated cases. Name the input files (respectively):

1Element_TXC_CamClay_NC.inp,

1Element_TXC_CamClay_LightOCR.inp,

1Element_TXC_CamClay_HeavyOCR.inp

```
*NODE, NSET=AllNodes
1, 0., 0., 0.
2, 1., 0., 0.
3, 1., 1., 0.
4, 0., 1., 0.
5, 0., 0., 1.
6, 1., 0., 1.
7, 1., 1., 1.
8, 0., 1., 1.
*ELEMENT, TYPE=C3D8, ELSET=PSolid
1, 1, 2, 3, 4, 5, 6, 7, 8
*NSET,NSET=PosXNodes
2,3,6,7
*NSET,NSET=PosYNodes
3,4,7,8
*NSET,NSET=PosZNodes
5,6,7,8
*************************************************
*SURFACE, NAME=PosZSurf, TYPE=ELEMENT
 1, S2
*SURFACE, NAME=PosXSurf, TYPE=ELEMENT
 1, S4
*SURFACE, NAME=PosYSurf, TYPE=ELEMENT
```

```
  1, S5
**************************************************
*SOLID SECTION, ELSET=PSolid, MATERIAL=MSolid
**************************************************
***
*** Units:
***    mass   = kg
***    length = mm
***    time   = s
***    force  = mN
***    stress = 1.0e+03 Pa
***
*PARAMETER
nu = 0.3
Mfac = 1.2
Lambda = 0.077
Kappa  = 0.0066
e0     = 0.5
**************************************************
***
*** Normally consolidated
***
**************************************************
Pc0    = 200.
a0     = 0.5*Pc0
ConfinementPressure = Pc0
**************************************************
***
*** Lightly overconsolidated
***
**************************************************
*** Pc0    = 200.
*** a0     = 0.5*Pc0
*** ConfinementPressure = Pc0/2.
**************************************************
***
*** Heavily overconsolidated
***
**************************************************
*** Pc0    = 500.
*** a0     = 0.5*Pc0
*** ConfinementPressure = Pc0/5.
```

```
*************************************************
NegConfinementPressure = -ConfinementPressure
*************************************************
COMP_STRAIN = 0.2
STEP_VELOC = -COMP_STRAIN
*************************************************
** Material Cards
*************************************************
*MATERIAL, NAME=MSolid
*POROUS ELASTIC, SHEAR=POISSON, TYPE=LOGARITHMIC
<Kappa>,<nu>
*CLAY PLASTICITY, HARDENING=EXPONENTIAL
<Lambda>,<Mfac>,<a0>
*INITIAL CONDITIONS, TYPE=RATIO
AllNodes,<e0>,0.,<e0>,1.
*INITIAL CONDITIONS,TYPE=STRESS,GEOSTATIC
PSolid,<NegConfinementPressure>,0.,<NegConfinementPressure>,1.,1.
*************************************************
** Load Amplitudes
*************************************************
*AMPLITUDE, TIME=TOTAL TIME, NAME=ConfinementPressure_Amplitude
0.0,0.0
1.0,<ConfinementPressure>
2.0,<ConfinementPressure>
*************************************************
** Step 1 - Initial Confinement
*************************************************
*STEP, UNSYMM=YES, NLGEOM=NO, INC=999
Confinement by geostatic
*GEOSTATIC
0.1,1.0,1.e-12,0.1
***********************
** Boundary conditions
***********************
*BOUNDARY, OP=MOD
1, 1
1, 2
1, 3
2, 2
2, 3
3, 3
4, 1
```

```
4, 3
5, 1
5, 2
6, 2
8, 1
************************
** Confinement pressure
************************
*DSLOAD, AMPLITUDE=ConfinementPressure_Amplitude, OP=NEW
PosXSurf,P,1.0
PosYSurf,P,1.0
PosZSurf,P,1.0
*******************
** ODB output
*******************
*OUTPUT, FIELD, FREQUENCY=1
*NODE OUTPUT
U,RF
*ELEMENT OUTPUT
E,S,PEEQ
*******************
** History output
*******************
*OUTPUT, HISTORY, FREQUENCY=1
*ELEMENT OUTPUT, ELSET=PSolid
E11,E22,E33,S11,S22,S33,MISES,PRESS,PEEQ
*END STEP
*************************************************
** Step 2 - Triaxial Loading
*************************************************
*STEP, UNSYMM=YES, NLGEOM=NO, INC=99999999
Triaxial Compression
*STATIC
0.01,1.0,1.e-12,0.1
************************
** Boundary conditions
************************
*BOUNDARY, TYPE=VELOCITY, OP=NEW
1, 1
1, 2
1, 3
2, 2
```

```
2, 3
3, 3
4, 1
4, 3
5, 1
5, 2
6, 2
8, 1
PosZNodes,3,3,<STEP_VELOC>
************************
** Confinement pressure
************************
*DSLOAD, AMPLITUDE=ConfinementPressure_Amplitude, OP=NEW
PosXSurf,P,1.0
PosYSurf,P,1.0
*END STEP
```

3.3.2 Cam-Clay Plotting script

```
import matplotlib.pyplot as plt
import math
import numpy as np
plt.rcParams.update({'font.size': 12})

Mfac = 1.2
ocrVec = ['NC','LightOCR','HeavyOCR']
basefig_filename = 'qp_plotCC'

def voffset(arr):
    v0  = arr[0]
    arr = [val - v0 for val in arr]
    return arr
# end def

for ocr in ocrVec:

    fig = plt.figure(figsize=(5,10))
    fig.subplots_adjust(hspace=0.5)
    ax1 = fig.add_subplot(3,1,1)
    ax2 = fig.add_subplot(3,1,2)
    ax3 = fig.add_subplot(3,1,3)

    jobName = '1Element_TXC_CamClay_' + ocr
    fin = open('results_'+jobName+'.csv','r')
    lines = fin.readlines()
    fin.close()

    axial_strain      = []
    volumetric_strain = []
    Pvec              = []
    Qvec              = []
    maj_rad_vec       = []
    for i,line in enumerate(lines):
        if i > 0: # skip headers
            print(line)
            line = line.strip()
            line = line.split(',')
            EA = float(line[0])
            EV = float(line[1])
```

```python
        Q  = float(line[2])
        P  = float(line[3])
        aval = float(line[4])
        axial_strain.append(EA)
        volumetric_strain.append(EV)
        Pvec.append(P)
        Qvec.append(Q)
        maj_rad_vec.append(aval)
        #min_rad_vec.append(aval*Mfac)

    # end if
# end for

ax1.plot(axial_strain,Qvec,'-o',\
    label=ocr)

ax2.plot(axial_strain,volumetric_strain,'-o',\
    label=ocr)

for aval in maj_rad_vec:
    maj_radius = aval
    min_radius = Mfac*aval
    t = np.linspace(0,360,360)
    x = maj_radius*np.cos(np.radians(t))+aval
    y = min_radius*np.sin(np.radians(t))
    ax3.plot(x,y,'k--')
# end for

ax3.plot(Pvec,Qvec,'-o',\
    label=ocr)

ax1.legend(framealpha=0.0, loc='center right')
ax2.legend(framealpha=0.0, loc='best')
ax3.legend(framealpha=0.0, loc='upper right')

ax1.set_xlabel('Axial Strain $\epsilon_a$ (%)')
ax1.set_ylabel('Shear $q$ (kPa)')
ax2.set_xlabel('Axial Strain $\epsilon_a$ (%)')
ax2.set_ylabel('Volumetric Strain $\epsilon_v$ (%)')

ax3.set_xlabel('Pressure $p\'$ (kPa)')
ax3.set_ylabel('Shear $q$ (kPa)')
```

```
    ax3.set_xlim([0,700])
    ax3.set_ylim([0,700])
    plt.tight_layout()
    plt.savefig(basefig_filename+'_'+ocr+'.png', dpi=300)
    #plt.show()

# end for
```

4 Slope Stability Analysis

Technically speaking, the proper way to obtain a global factor of safety (FS) is by means of limit equilibrium analysis (e.g., slip circle method) or more rigorously, by limit analysis. In the latter, rigorous plasticity theory is used to directly determine estimates of the collapse load of a given structural model without resorting to iterative or incremental analysis.

From a practical standpoint, however, it is difficult to use limit equilibrium analysis or limit analysis for geomechanical and geotechnical problems that have complicated geometries (which may include other structural elements), loads and materials. For this reason, various methods have been proposed to estimate the FS through finite element analysis. The most popular approach is the so-called strength reduction method (SRM) [2]. SRM is used with the Mohr-Coulomb material model in which the cohesion and friction angle are changed in each iteration i as follows:

$$c_i = \frac{c}{FS} \tag{26}$$

$$\tan(\phi_i) = \frac{\tan(\phi)}{FS} \tag{27}$$

$$\tan(\psi_i) = \frac{\tan(\psi)}{FS} \tag{28}$$

Essentially, SRM is a procedure in which the FS is obtained by weakening the soil until a global instability (slope failure) is obtained.

The most important element for the successful application of SRM is the criterion to detect global instability. Although non-convergence is often taken as a criterion for detecting global instability, the application of this criterion is by no means straightforward for a general nonlinear finite element model. For e.g., local non-convergence often prematurely terminate a finite element analysis, due to local or numerical instabilities when using non-associated plasticity [3]. The presence of other structural elements (which could also respond nonlinearly), as well as complicated loads and geometries can exacerbate this problem.

Abaqus has no automated way of determining the FS slope stability. Section 4.1.2 provides a Python script to iteratively determine the FS for the example problem shown in Figure8; the base (parameterized) input file given in Section 4.1.1. The SRM procedure is:

1. Start with some initial factor of safety interval $[FS_{left}, FS_{right}]$, with FS_{left} corresponding to a globally stable model and FS_{right} corresponding to a model that is globally unstable.

2. In each iteration i, run model with reduced soil cohesion, friction angle and dilation angle for the Mohr-Coulomb material using Equations (26), (27) and (28), where $FS = \frac{1}{2}(FS_{left} + FS_{right})$.

   The Python script parses the base input file and replaces the tag `<val>` on the line `FS = <val>` with a new FS value. A modified input file is then generated, and the corresponding Abaqus job is executed through the `os.system(...)` call from within the Python script.

3. Check whether global instability criterion is met. The global instability detection criterion is defined in the function `check_status`. Without attempting to cover all possibilities that can lead to a global instability, and to keep the implementation simple for this example, we use an ad-hoc rule for based on the ratio of the total model plastic dissipated energy to the total model internal strain energy. If this ratio is greater than 0.2, the model is considered globally unstable.

   The user use other methods for global instability detection, for e.g., tracking the maximum displacement as a function of FS [2]. As previously mentioned, non-convergence may not be a suitable criterion for detecting global instability due to the numerical or local instabilities in nonlinear finite element models.

4. Reset FS interval:

$$[FS_{left}, FS_{right}] := \begin{cases} [FS_{left}, FS] & \text{if model is unstable} \\ [FS, FS_{right}] & \text{otherwise} \end{cases}$$

5. Proceed with iteration until $\Delta FS = FS_{right} - FS_{left}$ is smaller than some prescribed tolerance.

4.1 Example

For the example shown in Figure8, a prescribed tolerance of $\Delta FS = 0.01$ is specified. The change in the FS interval with iteration is shown in Figure 9. The converged FS interval is $[1.106250, 1.115625]$. The corresponding equivalent plastic strain and displacement contours are shown in Figures 10a and 10b.

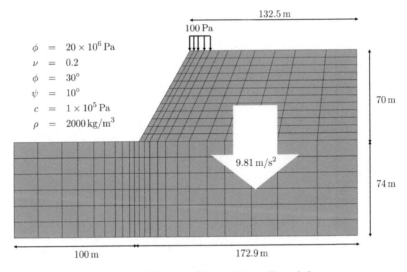

Figure 8: Slope Stability Problem.

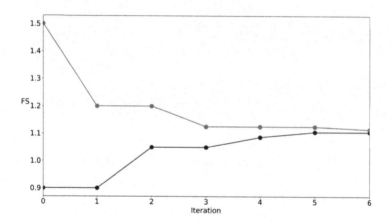

Figure 9: FS Interval History.

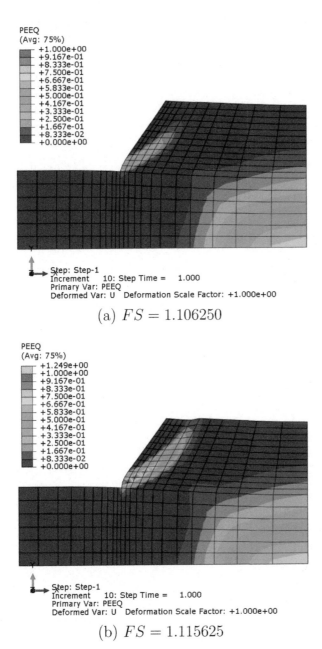

(a) $FS = 1.106250$

(b) $FS = 1.115625$

Figure 10: Contours of Equivalent Plastic Strains and Deformed Shapes.

4.1.1   Base Input File for Slope stability

```
*NODE
***
*** Y = 0
***
1,0.,0.
11,100.,0.
23,272.9,0.
***
*** Y = 74
***
241,0.,74.
251,100.,74.
263,272.9,74.
***
*** Y = 144
***
731,140.4,144.
743,272.9,144.
**
** bottom left boundary
**
*NGEN,NSET=BLHS
1,241,40
**
** vertical plane
**
*NGEN,NSET=BCEN
11,251,40
**
** bottom right boundary
**
*NGEN,NSET=BRHS
23,263,40
**
** slope
**
*NGEN,NSET=TCEN
251,731,40
```

```
**
** top right boundary
**
*NGEN,NSET=TRHS
263,743,40
*NFILL,BIAS=1.5,TWO STEP
BLHS,BCEN,10
*NFILL,BIAS=.66666666,TWO STEP
BCEN,BRHS,12
*NFILL,BIAS=.66666666,TWO STEP
TCEN,TRHS,12
*NSET,NSET=SLHS,GENERATE
241,251
*NSET,NSET=SRHS,GENERATE
731,743
*NSET,NSET=BOT,GENERATE
1,23
*NSET,NSET=FILN
251,411,731
*ELEMENT,TYPE=CPE4
1,1,2,42,41
101,11,12,52,51
*ELGEN,ELSET=ALLE
1,6,40,1,10,1,10
101,18,40,1,12,1,20
**************************************************
***
*** Units:
***    mass   = kg
***    length = m
***    time   = s
***    force  = N
***    stress = Pa
***
***
*************************
*PARAMETER
Eyoung = 20.e6
nu = 0.2
phi = 30.
psi = 10.
coh = 1.e5
```

```
rho = 2000.
***
FS = <val>
pival = 4.0*atan(1.0)
coh_adj = coh/FS
phi_adj = atan( tan(phi*pival/180.)/FS ) * 180./pival
psi_adj = atan( tan(psi*pival/180.)/FS ) * 180./pival
***
monitor_node = 331
monitor_dof  = 1
*************************
*SOLID SECTION,ELSET=ALLE,MATERIAL=ALLE
*MATERIAL,NAME=ALLE
*ELASTIC
<Eyoung>,<nu>
*MOHR COULOMB
<phi_adj>,<psi_adj>
*MOHR COULOMB HARDENING
<coh_adj>
*DENSITY
<rho>
***************************************************
*AMPLITUDE, NAME=RAMP, DEFINITION=SMOOTH STEP
0.,0., 1.,1.
***************************************************
*SURFACE, NAME=LoadSurf, TYPE=ELEMENT
118,S3
138,S3
158,S3
178,S3
198,S3
***************************************************
*STEP,UNSYMM=YES
*STATIC
0.1,1.0,1.e-6,0.1
*CONTROLS,ANALYSIS=DISCONTINUOUS
**********************
*BOUNDARY,OP=NEW
BOT,2,2,0.
BLHS,1,1,0.
BRHS,1,1,0.
TRHS,1,1,0.
```

```
**********************
*DSLOAD,AMPLITUDE=RAMP
LoadSurf,P,100.
*DLOAD,AMPLITUDE=RAMP
,GRAV,9.81,0.,-1.,0.
**********************
*OUTPUT,FIELD,FREQ=1
*NODE OUTPUT
U,
*ELEMENT OUTPUT
S,E,PEEQ
******************
*OUTPUT,HISTORY
*ENERGY OUTPUT, VARIABLES=PRESELECT
*NODE OUTPUT,NSET=FILN
U,
**********************
*MONITOR,NODE=<monitor_node>,DOF=<monitor_dof>
*END STEP
```

4.1.2 Abaqus Python File for Strength Reduction Method

This input file must be run using the Abaqus Python interface. Save the following lines as **run.py**, and then on the command prompt type:

abaqus python run.py

User-defined criterion to detect global instability is defined in the function **check_status**.

```
# ======================================
# must run this using Abaqus' Python !
# ======================================
from odbAccess import *
from odbMaterial import *
from odbSection import *
from abaqusConstants import *
import numpy as np
import os

# first run to start
ncpus         = 2
basefilename = 'base_model'
njters        = 20 # max number of jterations
FS_interval  = [0.9,1.5]
dFS_TOL       = 0.01 # stopping increment size
FS_left_history = [FS_interval[0]]
FS_right_history = [FS_interval[1]]

# ======================================
# helper functions
# ======================================
def RepresentsInt(s):
    try:
        int(s)
        return True
```

```python
        except ValueError:
            return False

# end def

def write_inpfile(jobname,jter,FSval,interval_end):

    #
    # read template inp file
    #
    fin = open( jobname + '.inp' , 'r' )
    lines = fin.readlines()
    fin.close()

    #
    # write inp file with FS value
    #
    newjobname = jobname + '_' + str(jter) + '_' + interval_end
    fout = open( newjobname + '.inp', 'w' )
    for line in lines:
        if 'FS = <val>' in line:
            line = 'FS = ' + str(FSval) + '\n'
        # end if
        fout.write(line)

    # end for
    fout.close()

    return newjobname

# end def

def write_inpfile_stabilized(basefilename,jobname,FSval):

    #
    # read template inp file
    #
    fin = open( basefilename + '.inp' , 'r' )
    lines = fin.readlines()
    fin.close()

    #
```

68

```python
    # write inp file with FS value
    #
    fout = open( jobname + '.inp', 'w' )
    for line in lines:
        if 'FS = <val>' in line:
            line = 'FS = ' + str(FSval) + '\n'
        # end if
        if '*STATIC' in line:
            line = '*STATIC,STABILIZE\n'
        # end if
        fout.write(line)
    # end for
    fout.close()


# end def

def check_status(jobname):

    status = 'PASS'

    # ------------------------------------------------
    # define problem-specific
    # user-criteria here, below is an example
    # ------------------------------------------------

    #
    # check energy from odb
    #
    odbName=jobname+'.odb'

    odb = openOdb(odbName)
    historyRegions = \
        odb.steps['Step-1'].\
        historyRegions['Assembly Assembly-1'].\
        historyOutputs

    ALLSE = historyRegions['ALLSE'].data[-1][1]
    ALLPD = historyRegions['ALLPD'].data[-1][1]

    odb.close()
```

```python
        if ALLPD/ALLSE > 0.2:
            status = 'FAIL'
        # end if


        #
        # check displacement magnitude
        # and/or convergence from sta file
        #
        fin = open( jobname + '.sta' , 'r' )
        lines = fin.readlines()
        fin.close()
        for line in lines:
            line = line.strip()
            if line: # if not empty line
                #line = line.split()
                # line with step number
                #if RepresentsInt(line[0]):
                #    monitor_dof = float(line[-1])
                #    if abs(monitor_dof) > 20.0:
                #        status = 'FAIL'
                #    # end if
                ## end if
                if 'THE ANALYSIS HAS NOT BEEN COMPLETED' in line:
                    status = 'FAIL'
                # end if
            # end if

        # end for



        return status

# end def

# =====================================
# iteration 0
# =====================================
jter = 0
#
# run for FS_left and FS_right
#
```

```
FS_left  = FS_interval[0]
FS_right = FS_interval[1]
print('=====================================')
print('  ITERATION 0 ')
print('  FS_left = %f, FS_right = %f '%(FS_left,FS_right))
print('=====================================')

jobname_left  = write_inpfile(basefilename,jter,FS_left,'left')
jobname_right = write_inpfile(basefilename,jter,FS_right,'right')
pass_model_name = jobname_left
fail_model_name = jobname_right
cmd_left  = 'abaqus job=' + jobname_left  \
    + ' cpus=' + str(ncpus) \
    + ' interactive ask_delete=off'
cmd_right = 'abaqus job=' + jobname_right \
    + ' cpus=' + str(ncpus) + \
    ' interactive ask_delete=off'
os.system(cmd_left)
os.system(cmd_right)
#
# check sta file for pass or fail
#
status_left = check_status(jobname_left)
status_right = check_status(jobname_right)

# =====================================
# check validity of initial FS interval
# =====================================
flag1 = 'FAIL' in status_left and 'PASS' in status_right
flag2 = 'FAIL' in status_left and 'FAIL' in status_right
if flag1 or flag2:
    raise Exception('left should be PASS and right should be FAIL')
# end if

# =====================================
# iteration 1 and beyond
# =====================================
complete_flag = False

for jter in range(1,njters+1):

    #
```

```python
# run model at FS_mid
#
FS_mid = 0.5*(FS_left+FS_right)
jobname_mid  = write_inpfile(basefilename,jter,FS_mid,'mid')
cmd_mid  = 'abaqus job=' \
    + jobname_mid \
    + ' interactive ask_delete=off'
os.system(cmd_mid)


#
# check sta file for pass or fail
#
status_mid = check_status(jobname_mid)


#
# recalculate FS_interval
#
if 'FAIL' in status_mid:
    FS_interval[1] = FS_mid
    fail_model_name = jobname_mid
else:
    FS_interval[0] = FS_mid
    pass_model_name = jobname_mid
# end if

FS_left  = FS_interval[0]
FS_right = FS_interval[1]

print('====================================')
print('  Completed ITERATION %d '%(jter))
print('  FS_left = %f, FS_right = %f '%(FS_left,FS_right))
print('====================================')

FS_left_history.append(FS_left)
FS_right_history.append(FS_right)

#
# check interval size
#
dFS = FS_right - FS_left

if dFS < dFS_TOL:
```

```python
            complete_flag = True
            break
        # end if


    print('*****************************************')
# end for

# ======================================
# check whether procedure is completed
# ======================================
if complete_flag:
    print('Procedure completed at iteration %d\n'%\
        (jter-1))
    print('Factor of safety is %f\n'%\
        (FS_interval[0]))
    print('Model that pass at FOS %f = %s\n'%\
        (FS_interval[0],pass_model_name))
    print('Model that fails at FOS %f = %s\n'%\
        (FS_interval[1],fail_model_name))

    #
    # if necessary, run a stabilized version of
    # the failed model for visualization purposes
    #
    #stabilized_failed_jobname = \
    #    fail_model_name+'_stabilized'
    #write_inpfile_stabilized(basefilename,\
    #    stabilized_failed_jobname,FS_interval[1])
    #cmd  = 'abaqus job=' + stabilized_failed_jobname\
    #    + ' interactive ask_delete=off'
    #os.system(cmd)

else:
    print('Procedure did not completed \
        in %d jterations\n'%(njters))
# end if

# ======================================
# write FS_interval_history
# ======================================
fout = open('FS_interval_history.csv','w')
```

73

```python
for j in range(0,len(FS_left_history)):
    FS_left = FS_left_history[j]
    FS_right = FS_right_history[j]
    fout.write('%d,%f,%f\n'%(j,FS_left,FS_right))
# end for
fout.close()
```

5 Consolidation

5.1 Dimensional Analysis

Notation for dimensional analysis:

- Length = L
- Time = T
- Mass = M
- Force = F = MLT^{-2}
- Stress = S = FL^{-2} = $ML^{-1}T^{-2}$

5.2 Forchheimer's law

Permeability in Abaqus is defined in general by Forchheimer's law, which accounts for changes in permeability as a function of fluid flow velocity $\mathbf{v}_w$:

$$\mathbf{f}\left(1 + \beta\sqrt{\mathbf{v}_w \cdot \mathbf{v}_w}\right) = -\frac{k_s}{\gamma_w}\mathbf{k} \cdot \left(\frac{\partial u_w}{\partial \mathbf{x}} - \rho_w \mathbf{g}\right) \quad (29)$$

where

- $\mathbf{f} = s\,n\,\mathbf{v}_w$ is the fluid flux $\left[LT^{-1}\right]$
- n is the porosity
- s is the fluid saturation
- u_w is the wetting fluid pore pressure

- $k_s = k_s(s)$ is the dependence of permeability on saturation of the wetting liquid such that $k_s = 1$ and $s = 1$

- γ_w is the specific weight (or unit weight) of the wetting fluid

- $\rho_w = \gamma_w/g$ is the mass density of the wetting fluid $\left[ML^{-3}\right]$

- g is the magnitude of gravitational acceleration $\left[LT^{-2}\right]$

- $\mathbf{k}$ is the (hydraulic) permeability tensor $\left[LT^{-1}\right]$

- $e = \dfrac{n}{1-n}$ is the void ratio

- $\beta = \beta(e)$ is a "velocity coefficient" which may be dependent on the void ratio of the material

5.3 Darcy's Law

The fluid flow model used many engineering type applications is based on Darcy's law in which the fluid velocity is "slow". Darcy's law can be obtained from Forchheimer's law by ignoring the velocity-dependent term, i.e., set $\beta = 0$:

$$
\begin{aligned}
\mathbf{f} &= -\frac{k_s}{\gamma_w}\mathbf{k} \cdot \left(\frac{\partial u_w}{\partial \mathbf{x}} - \rho_w \mathbf{g}\right) \\
&= -k_s \mathbf{k} \cdot \left(\frac{1}{\gamma_w}\frac{\partial u_w}{\partial \mathbf{x}} - \frac{\rho_w}{\gamma_w}\mathbf{g}\right) \\
&= -k_s \mathbf{k} \cdot \left(\frac{1}{\gamma_w}\frac{\partial u_w}{\partial \mathbf{x}} - \frac{1}{g}\mathbf{g}\right)
\end{aligned}
\tag{30}
$$

Furthermore, if we ignore the gravity term, assume isotropic (hydraulic) permeability, $\mathbf{k} = \bar{k}\,\mathbf{I}$ and fully saturated condition $s = 1$ (and therefore $k_s = 1$ and $\mathbf{f} = n\,\mathbf{v}_w$), the above simplifies as:

$$
\begin{aligned}
n\,\mathbf{v}_w &= -\frac{1}{\gamma_w}\bar{k}\,\mathbf{I}\cdot\left(\frac{\partial u_w}{\partial \mathbf{x}}\right) \\
&= -\left(\frac{\bar{k}}{\gamma_w}\right)\frac{\partial u_w}{\partial \mathbf{x}}
\end{aligned}
\tag{31}
$$

Note that both sides of the above have dimensions of velocity $\left[\mathrm{LT}^{-1}\right]$. In many books, the above form of the Darcy's Law is expressed as:

$$
\mathbf{q} = -\bar{k}\,\nabla h \tag{32}
$$

where:

- $\mathbf{q} = n\,\mathbf{v}_w$ is the fluid flux

- $\nabla h = \dfrac{1}{\gamma_w}\dfrac{\partial u_w}{\partial \mathbf{x}} = \dfrac{\partial\left(u_w/\gamma_w\right)}{\partial \mathbf{x}}$ (assuming spatially constant fluid specific weight) is the spatial gradient of the potential head h

$\bar{k}$ is often called the *hydraulic conductivity*.

5.4 Intrinsic Permeability

Sometimes, the so-called *intrinsic* permeability $\hat{k}$, which has dimensions of $[L^2]$ is used. The relationship between the intrinsic and hydraulic permeability is:

$$\bar{k} = \frac{\hat{k}\, g}{\eta_{kinematic}} \tag{33}$$

where $\mu_{dynamic}$ is the dynamic viscosity of the fluid and $\nu_{kinematic}$ is the kinematic viscosity of the fluid, with the relationship:

$$\nu_{kinematic} = \frac{\mu_{dynamic}}{\rho_w} \tag{34}$$

The dimensions of the types of viscosities:

- dynamic viscosity $= \mathrm{FTL}^{-2}$ or ST or $\mathrm{MT}^{-1}\mathrm{L}^{-1}$

- kinematic viscosity $= \mathrm{L}^2\mathrm{T}^{-1}$

A widely used model for the intrinsic permeability is the Carman-Kozeny model [4]:

$$\hat{k} = \frac{r_f^2}{4\, k_{kc}} \frac{n^3}{(1-n)^2} \tag{35}$$

where k_{kc} represents the Carman-Kozeny constant (parameter that is geometry dependent) and r_f represents the average radius of the porous particles/fibers.

5.5 Notes on Abaqus Coupled Pore Fluid and Stress Analysis

1. The *DENSITY keyword refers to the *dry* density of the soil material.

2. The initial void ratio e_0 is specified as an *initial condition* using the keyword
   *INITIAL CONDITIONS, TYPE=RATIO.

3. The coupled pore fluid diffusion/stress analysis step is invoked using the
   *SOILS,CONSOLIDATION,UTOL=<Utol>,END=PERIOD
   keyword. The UTOL=<Utol> requests that the time step is restricted to ensure that the maximum pore pressure change is limited by <Utol>. For transient analysis, the option END=PERIOD requires that a consolidation time period be specified by the user. If the user wants the analysis to be automatically terminated when a steady-state condition is reached, use END=SS instead.

4. The initial effective stress field (stresses are effective stresses when the analysis includes pore fluid flow) is established through a geostatic step using the keyword
   *GEOSTATIC. The initial conditions for effective stresses must be provided using the keyword
   *INITIAL CONDITIONS, TYPE=STRESS. The initial state of stress must be close to being in equilibrium with the applied loads and boundary conditions.

   Alternatively, for problems with simple geometries and loads (such as the example shown in Figure 11), the initial effective stress may be directly established using a regular consolidation step *SOILS,CONSOLIDATION.

5. If the magnitude and direction of the gravitational loading are defined by using the gravity distributed load type (GRAV load type under *DLOAD), a total pore pressure solution is used. Excess pore pressure solutions are provided in all other cases; for example, when gravity loading is defined with body force distributed loads (BX, BY or BZ load type under *DLOAD).

6. The specific weight of the fluid γ_w *must be specified correctly* even if the analysis does not consider the weight of the wetting liquid (i.e., if excess pore fluid pressure is calculated).

7. Spurious oscillations may appear in the solution when *small* time increments are used. A minimum time step is needed for solution stability. In the Abaqus documentation [5], the following guideline for minimum usable time increment is provided for fully saturated flow:

$$\Delta t > \frac{\gamma_w \left(1 + \beta v_w\right)}{6E\mathrm{k}} \left(1 - \frac{E}{K_g}\right)^2 (\Delta \ell)^2 \qquad (36)$$

5.6 Example

Consider the following column consolidation problem:

- Column height $= 4$ m (see Figure 11)

- Soil elastic modulus $E = 70 \times 10^6$ Pa

- Soil Poisson's ratio $\nu = 0$

- Soil porosity $n = 0.3$

- Carman-Kozeny constant $k_{kc} = 45$

- Mean grain size $r_f = 1 \times 10^{-4}$ m

- Fluid (water) density $\rho_w = 1000$ kg/m^3

- Dynamic viscosity of fluid $\mu_{dynamic} = 1 \times 10^{-3}$ Pa-sec

- A load of 90 kPa is suddenly applied at the top of the column.

The input file is given in Section 5.6.1. The resulting pore pressure profiles are shown in Figure 12. The pore pressure starts off as constant at 90 kPa (rightmost vertical line in the figure). When drainage is allowed at the top, the pore pressure becomes zero there. The pore pressure then dissipates with time (profiles shifting to the left of the figure).

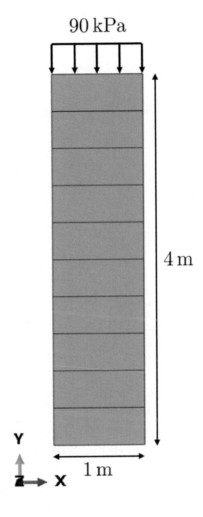

Figure 11: Consolidation Problem.

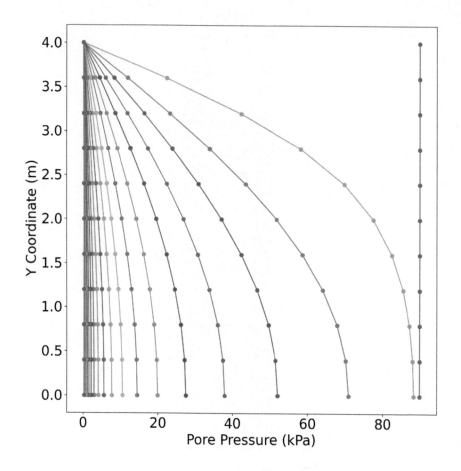

Figure 12: Change in Pore Pressure Profile with Time.

5.6.1   Input File for Consolidation Example

Save the following lines as `consolidation.inp`, and then on the command prompt type:

`abaqus job=consolidation interactive`

```
*PARAMETER
***
*** column height
***
height   = 4.
***
*** elastic properties
***
Ebar     = 70.e6
nu       = 0.
***
*** for permeability
***
grav       = 9.81
nporo      = 0.3
dmean      = 1.e-4
density_w  = 1000.
***
*** applied pressure
***
sigbar   = 90.e3
***
*** specific weight, void ratio
***
specific_weight_w = density_w*grav
evoid = nporo/(1.0-nporo)
***
*** intrinsic permeability by Kozeny-Carman, m^2
*** note: this is \hat{K} in Abaqus documentation
***
kkc = 45.
perm_intrinsic = (dmean**2/(4.*kkc))*nporo**3/(1-nporo)**2
***
```

```
*** Dynamic viscosity, N-sec/m^2 == Pa-sec
***
visc_dynamic    = 1.e-3
***
*** Kinematic viscosity, m^2/sec
***
visc_kinematic = visc_dynamic/density_w
***
*** hydraulic permeability
*** note: this is \bar{k} in Abaqus documentation
***
perm_hydraulic = perm_intrinsic*grav/visc_kinematic
***
*** time stepping parameters
***
dt_min = 0.5
Utol = 5.e7
*************************
*NODE
1,0.0,0.0
2,0.5,0.0
3,1.0,0.0
201,0.0,<height>
202,0.5,<height>
203,1.0,<height>
*NGEN,NSET=NALL
1,201,10
2,202,20
3,203,10
*NSET,NSET=FILE
141,
*NSET,NSET=TOP
201,202,203
*NSET,NSET=BASE
1,2,3
*NSET,NSET=NCENTER,GENERATE
2,202,20
*************************
*ELEMENT,TYPE=CPE8P,ELSET=ONE
1,1,3,23,21,2,13,22,11
*ELGEN,ELSET=ALL
1,10,20
```

```
*ELSET,ELSET=P1
6,7,8
***************************
*SOLID SECTION,MATERIAL=A1,ELSET=ALL
*MATERIAL,NAME=A1
*ELASTIC
<Ebar>,<nu>
*PERMEABILITY,SPECIFIC=<specific_weight_w>
<perm_hydraulic>,
*INITIAL CONDITIONS,TYPE=RATIO
NALL,<evoid>,0.,<evoid>,<height>
***************************
***
*** Boundary conditions on solid
***
*BOUNDARY
BASE,1,2
NALL,1
***
*** Boundary conditions on pore fluid
*** is initially undrained
***
***************************
***************************
***************************
*STEP
SUDDENLY APPLIED LOAD
*SOILS,CONSOLIDATION
1.E-7,1.E-7
****
*DLOAD
10,P3,<sigbar>
****
*OUTPUT,FIELD,FREQ=1
*NODE OUTPUT
U,POR,RVT
*ELEMENT OUTPUT
E,S
*OUTPUT,HISTORY,FREQ=1
*END STEP
***************************
***************************
```

86

```
***************************
*STEP,INC=999999999
 CONSOLIDATE
*SOILS,CONSOLIDATION,UTOL=<Utol>,END=PERIOD
<dt_min>,200.,<dt_min>,2.
*BOUNDARY
TOP,8
*OUTPUT,FIELD
*ELEMENT OUTPUT
E,S
*NODE OUTPUT
U,POR
*END STEP
```

5.6.2 Abaqus Python Post Processing File

Save the following lines as **postprocess.py**, and then on the command prompt type:

abaqus python postprocess.py

```
from odbAccess import *
from odbMaterial import *
from odbSection import *
from abaqusConstants import *
import numpy as np

jobNameVec = ['consolidation']
for jobName in jobNameVec:

    odbName=jobName+'.odb'

    odb = openOdb(odbName)
    assembly = odb.rootAssembly

    #
    # nodes at X = 0.5
    #
    nodeAllSet = \
        odb.rootAssembly.\
        instances['PART-1-1'].\
        nodeSets['NALL']

    #
    # build nodal coordinates dictionary
    #
    node_dict = {}
    for node in nodeAllSet.nodes:
        nodeLabel = node.label
        node_dict[nodeLabel] = node.coordinates.tolist()
    # end for

    nodeCenterSet = \
        odb.rootAssembly.\
```

```
                  instances['PART-1-1'].nodeSets['NCENTER']

    print('Processing Step: %d\n'%\
          odb.steps['Step-2'].number)

    frames = odb.steps['Step-2'].frames
    for index in range(0,len(frames)):

        frame = frames[index]
        print('Frame: %d'% index)

        PORSet = frame.fieldOutputs['POR'].\
            getSubset(region=nodeCenterSet)

        PORVals = PORSet.values

        #
        # write out results for this frame
        #
        f = open('results_'+jobName+'_frame'+str(index)+'.csv', 'w')
        for j in range(0,len(PORVals)):
            nodeLabel  = PORVals[j].nodeLabel
            nodeCoord  = node_dict[nodeLabel]
            nodeYCoord = nodeCoord[1]
            nodePORVal = PORVals[j].data
            f.write('%20.8e,%20.8e\n'\
                %(nodeYCoord,nodePORVal))
        # end for
        f.close()

    # end for

    # Close the output database before exiting the program
    odb.close()

# end for
```

5.6.3 Python Plotting File

Note: the following script requires a separate Python installation that includes the Matplotlib plotting library. Save the following lines as **plot_results.py**, and then on the command prompt type:

python plot_results.py

```python
import matplotlib.pyplot as plt
import math
import numpy as np
plt.rcParams.update({'font.size': 22})

base_jobname = 'consolidation'
figfilename = base_jobname + '_history.png'
nframes=103
skip=5

fig = plt.figure(figsize=(10,10))
#fig.subplots_adjust(hspace=0.5)
#ax1 = fig.add_subplot(3,1,1)
#ax2 = fig.add_subplot(3,1,2)
#ax3 = fig.add_subplot(3,1,3)


for index in range(0,nframes,skip):

    csvFile = 'results_' + \
        base_jobname + \
        '_frame' + \
        str(index) + '.csv'

    fin = open(csvFile,'r')
    lines = fin.readlines()
    fin.close()

    Yvec = []
    PORvec = []
```

```python
    for i,line in enumerate(lines):
        line = line.strip()
        line = line.split(',')
        print(line[0],line[1])
        YCoord = float(line[0])
        PORVal = float(line[1])/1000.
        Yvec.append(YCoord)
        PORvec.append(PORVal)
    # end for

    plt.plot(PORvec,Yvec,'-o')

# end for

plt.xlabel('Pore Pressure (kPa)')
plt.ylabel('Y Coordinate (m)')
plt.tight_layout()
plt.savefig(figfilename, dpi=300)
plt.show()
```

References

[1] D. Muir Wood. *Soil Behaviour and Critical State Soil Mechanics*. Cambridge University Press, Cambridge, UK, 1990.

[2] DV Griffiths and PA Lane. Slope stability analysis by finite elements. *Geotechnique*, 49(3):387–403, 1999.

[3] Franz Tschuchnigg, HF Schweiger, and Scott W Sloan. Slope stability analysis by means of finite element limit analysis and finite element strength reduction techniques. Part I: Numerical studies considering non-associated plasticity. *Computers and Geotechnics*, 70:169–177, 2015.

[4] Olivier Coussy. *Mechanics and physics of porous solids*. John Wiley & Sons, 2011.

[5] Abaqus Documentation, Dassault Systèmes support knowledge base. `https://support.3ds.com/knowledge-base/`.

Printed in Great Britain
by Amazon

40323494R00056